I0483526

Dugout Canoes in photos

Indigenous Technology Volume II

in color

Cedargrove Mastermind Group

ISBN-13: 978-1530062461

ISBN-10: 1530062462

Dugout Canoes in photos   2   ISBN-13: 978-1530062461   ISBN-10: 1530062462

Introduction

Dugout canoes are known worldwide, in any place where there are suitable trees, and bodies of water. The process of making them is not as simple as it seems. Hawaiians, in seeking out a tree to make dugout outrigger canoes, sought out trees that had no woodpeckers, or other indications of insects. This was one way to achieve quality control.

The process seems simple enough: fell the tree, pick the best side for the bottom, and build a fire on top. The fire would burn. The charred area would be removed, perhaps with stone tools, and a fire built again. The process took a while. In a way, the dugout canoe is a large, pointed bowl, of wood, in the same way that a birchbark canoe is a large, pointed basket.

Storage was easy enough. In the winter, the boats could be weighted with rocks, and stored at the bottom of a pond. Several dugouts like this have been found in Connecticut. Otherwise, we suspect that they were stored in shade, just as Hawaiian dugouts were, when not in use.

We are used to a high polish, smooth finish on wooden items. Hawaiian dugout outrigger canoes can be finished to a high polish, but this was probably not typical. They had sandpaper, of a sort, sand on treated leather. They could use oil, to finish the wood. When I made my first bow, out of Osage Orange, I heated the wood, over a fire, and put oil on the wood. The wood soaked it up right away. A living tree has a humidity of around 40%. When dried, wood has about 10% humidity. It soaks up oil, to replace the water. One can even use vegetable oil, for finishing wood.

Tom Brown, Jr., noted that a bow bought in a store is a piece of technology. A bow made with iron tools is a personal implement. A bow made with stone tools is an extension of your arm, which is one reason indigenous peoples buried their dead with their tools. A dugout canoe, made in the old way, is almost alive, and feels very different from, say, an aluminum canoe.

A picture is worth 1,000 words.

This was taken in a museum. This boat was built by an Ojibwa man, in the old way.

ISBN-13: 978-1530062461   ISBN-10: 1530062462

Some flaws in the wood appeared to have been repaired. Many museums will not allow flash photography, though in this one, they didn't mind. The wood has an odd color because of the flash.

This is a closeup of the wood, in the dugout.

This dugout canoe could hold more than one person.

You can see the interior of the dugout, here.

You can see the inside of the front of the canoe, with a shoe for size comparison. Note how thick the walls of the dugout are. They were about 3"/8 cm thick, over most of the length. I'm not sure what was used for the caulking.

This is a large dugout, of a sort known in New England. Making one of these is not as easy as it looks. The hull has to be thicker, so the dugout will float safely, without tipping over. The sides are flat. Creating this dugout took some careful work.

The line drawing makes the grain of the wood a little easer to see.

This is the dugout canoe from a distance. Perspective was a little difficult in this museum.

Dugout Canoes in photos　　　　　12　　　　ISBN-13: 978-1530062461　　ISBN-10: 1530062462

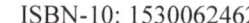

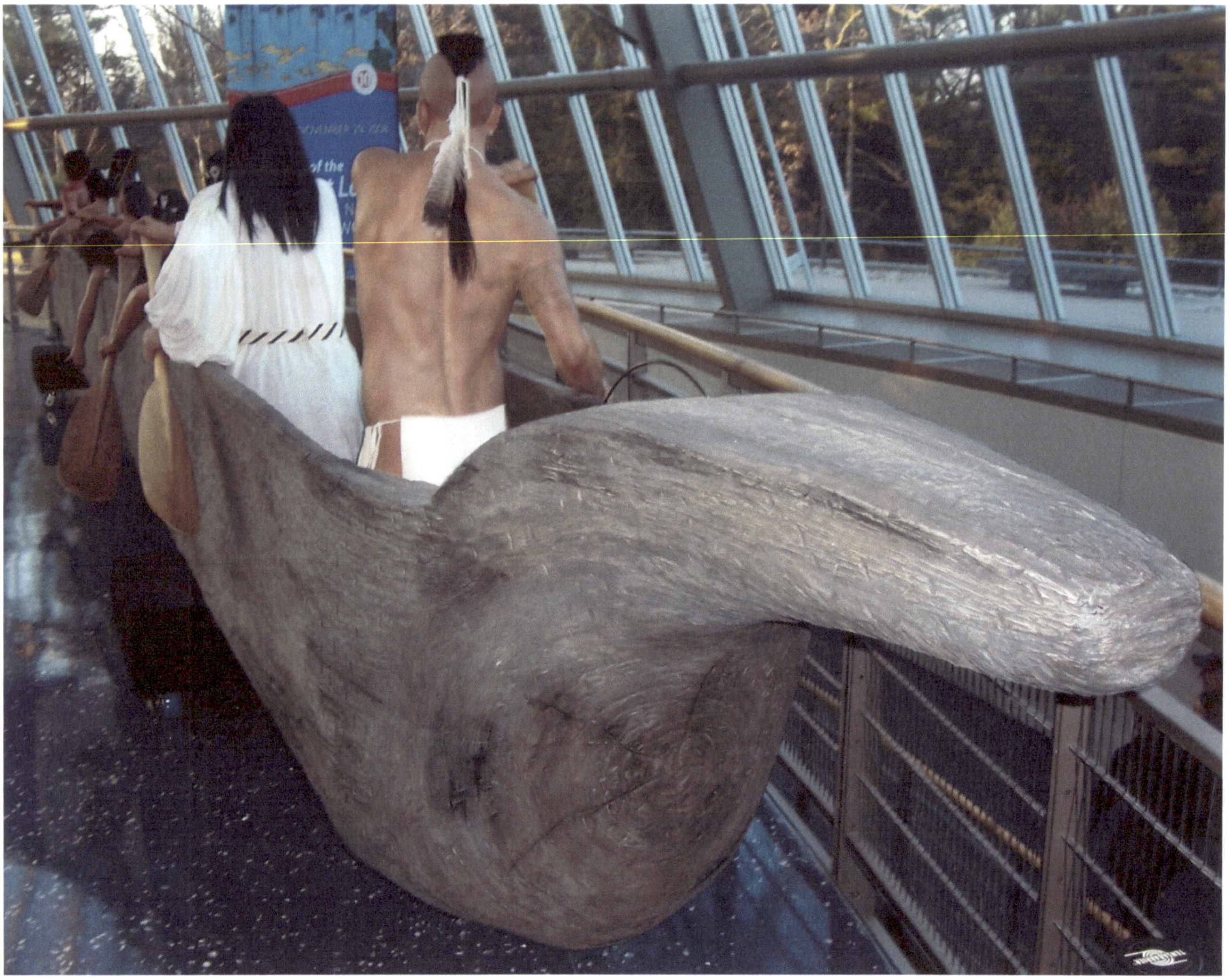

This is a rear view of the dugout, which also shows the grain of the wood. Note the tree ring.

These dugouts were at a museum of living history. Note the paddles at the top. The lower picture gives a good idea of shape.

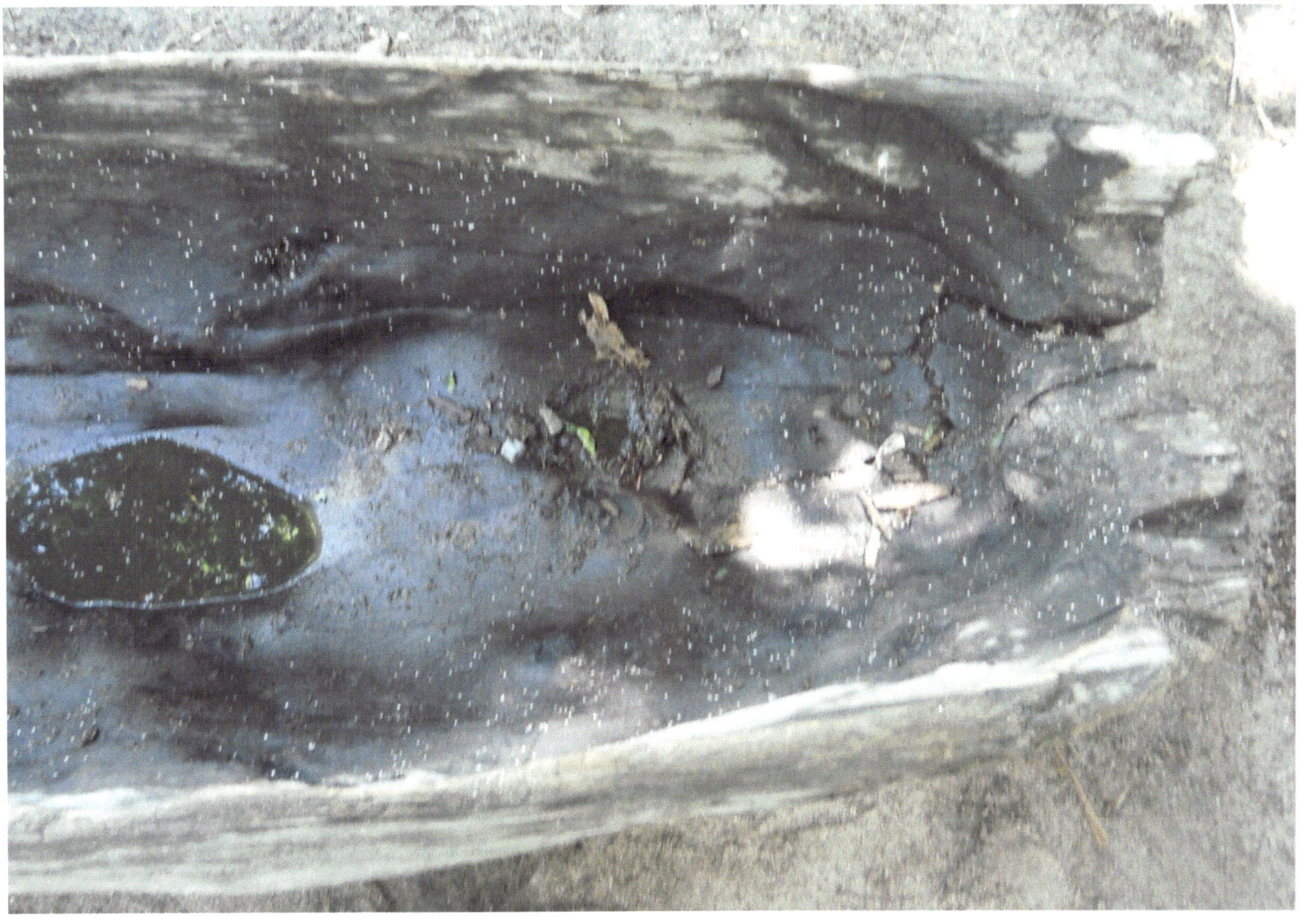

It had rained the previous day. Note that the dugout just above was kept in shade. The figure at left gives the size, of this dugout canoe. Note the shaded bower at right, for resting in the midday sun. The lake is just beyond.

ISBN-13: 978-1530062461   ISBN-10: 1530062462

These pictures give a good idea of the shape of a dugout canoe.

I'm not certain, however it appeared that these dugouts were in the final stages of being hollowed out, with fire, and tools to remove the charcoal.

Dugout Canoes in photos 18 ISBN-13: 978-1530062461 ISBN-10: 1530062462

You can see my shoe, to set size. Burning out a dugout canoe with fire doesn't usually give an even surface.

The shoes give the scale. Notice how uneven the bottom is, as shown by the water level. The ends of dugout canoes were thicker, in part because the grain of the wood was weaker there.

These are flatter dugouts, that were in the water. At this particular museum, I saw over 10 such flat dugout canoes, in the water, years ago. This was what they had on this day. Notice the rainwater in the dugout canoes.

Note the dugout that is almost submerged. This was one way to store a dugout.

ISBN-13: 978-1530062461   ISBN-10: 1530062462

You can clearly see the grain of the wood, in this picture.

Prior to the Civil War, in the USA, most household implements were made out of wood. Factories built to meet Civil War needs changed over to manufacturing civilian goods, and so more and more metal implements came to be made.

The cultures that made dugouts in the north tended to live in houses that looked something like this. What would it be like, to trust your life to wood, in this way? What would it be like, to be tactile with wood, for a good part of the day?

The above was in the Spanish Naval Museum, in Madrid. They had a number of boat models, of boats used in various Spanish colonial possessions. This is an outrigger dugout canoe, possibly from the Phillipines. They most absolutely forbade the use of a flash, but at least let me take pictures.

The drawing may make it a bit easier to see.

ISBN-13: 978-1530062461　ISBN-10: 1530062462

The Polynesians that populated many Pacific islands used sails, on their outrigger dugout canoes. This is shown in this model, at the Spanish Naval Museum, in Madrid.

Note the approximately ¼ scale outrigger dugout canoe, at the top.

ISBN-13: 978-1530062461 ISBN-10: 1530062462

The line drawing shows some details not obvious in the photographs.

ISBN-13: 978-1530062461   ISBN-10: 1530062462

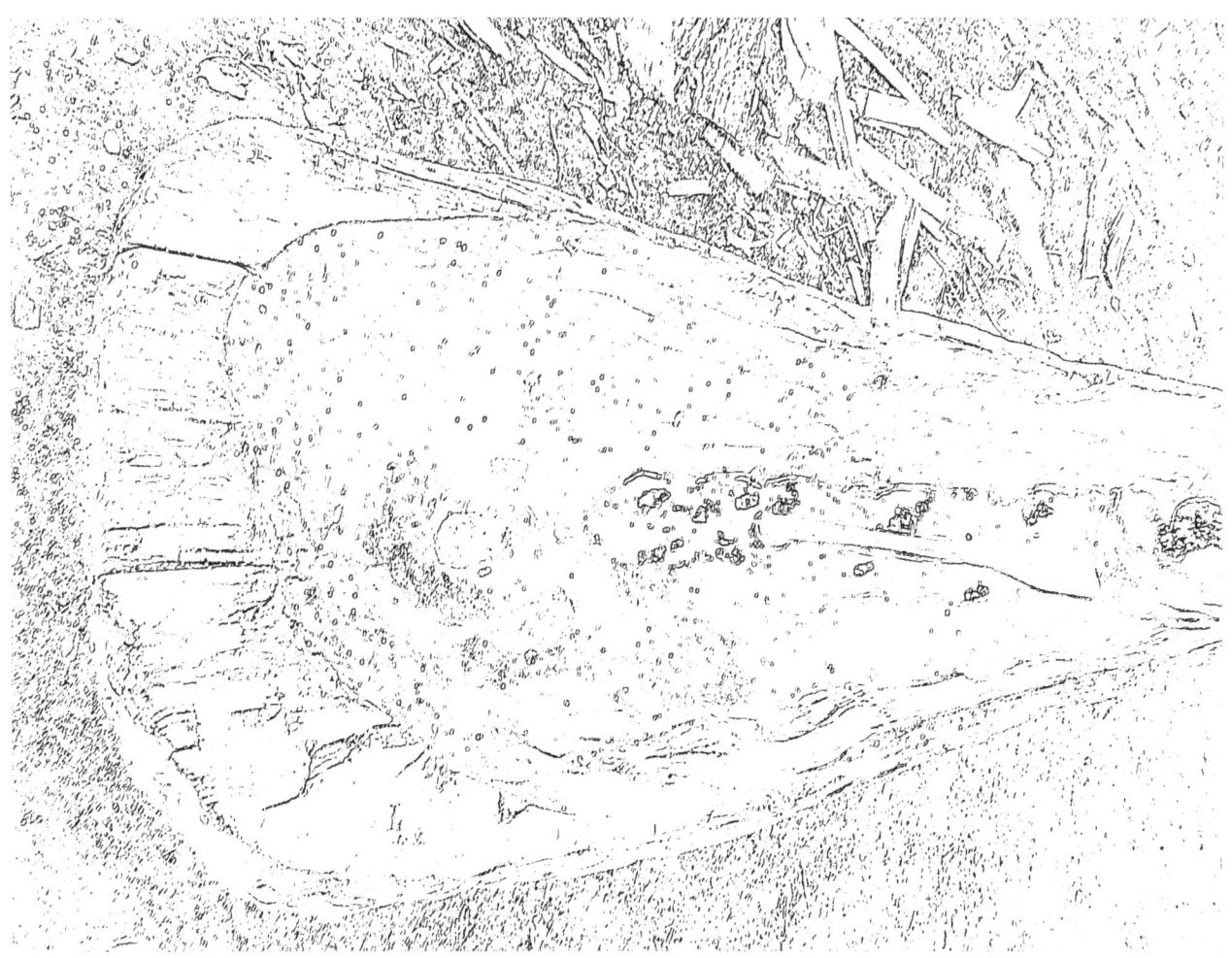

This kind of drawing helps me see detail in photographs.

www.ingramcontent.com/pod-product-compliance
Lightning Source LLC
Chambersburg PA
CBHW040756200526
45159CB00026B/2868